George Jacob Ziegler

Researches on the Medical Properties and Applications of

Nitrous Oxide

George Jacob Ziegler

Researches on the Medical Properties and Applications of Nitrous Oxide

ISBN/EAN: 9783337345747

Printed in Europe, USA, Canada, Australia, Japan

Cover: Foto ©berggeist007 / pixelio.de

More available books at **www.hansebooks.com**

RESEARCHES

ON THE

MEDICAL PROPERTIES AND APPLICATIONS

OF

NITROUS OXIDE,

PROTOXIDE OF NITROGEN,

OR LAUGHING GAS.

BY

GEO. J. ZIEGLER, M. D.,

PHYSICIAN TO THE PHILADELPHIA HOSPITAL, MEMBER OF THE AMERICAN MEDICAL
ASSOCIATION, MEMBER OF THE ACADEMY OF NATURAL SCIENCES
OF PHILADELPHIA, ETC. ETC.

Revised and Republished from the Med. and Surg. Reporter.

PHILADELPHIA:

J. B. LIPPINCOTT & CO.

1865.

PREFACE.

In the following pages, with some general reflections on collateral subjects, I have endeavored to present a condensed summary of the medical properties and applications of nitrous oxide as determined by a series of observations from time to time during a period of about sixteen years. These are necessarily incomplete and to a certain extent inconclusive, yet hoped they are sufficiently impressive to attract attention and induce more enlarged effort to determine the sanitive value of this remarkable agent and extend its sphere of usefulness in the preservation of life, promotion of health, and the relief of disease. Hence a close scrutiny of the facts presented and ideas advanced, and a liberal examination of the sanitive qualities of protoxide of nitrogen are particularly invited, believing that a more extended and thorough investigation of its properties will not only confirm the statements herein made and enable physicians to attain a greater degree of success in the amelioration of the ills of life, but will otherwise result in great good to the medical profession and humanity at large.   G. J. Z.

Philadelphia, July 15th, 1865.

MEDICAL PROPERTIES AND APPLICATIONS

OF

PROTOXIDE OF NITROGEN,

NITROUS OXIDE, OR LAUGHING GAS.

INTRODUCTORY.

NOTWITHSTANDING the present activity in the search for new remedies and the manifest desire to discover such as are both conservative in character and efficient in action, attention appears to be still too exclusively concentrated upon those agents which are more purely medicinal in their nature rather than upon such as have a direct physiological compatibility with the vital organism. The undue tendency in this direction has of recent years, however, been materially modified in consequence of a better appreciation of the general laws and correlations of physiology, pathology, hygiene, and therapeutics.

With this increased knowledge of medical science a more philosophical conception of the inherent nature and biological attributes of remedies has been acquired whereby it has become apparent that those agents which either subserve the immediate purpose of supplying

1*   (5)

elements of nutrition to, or in exerting a direct influence upon the functions of the living economy, or both combined, are, in the main, the most efficient and reliable for therapeutic purposes.* Hence, notwithstanding apparent exceptions, it may be regarded as a general truism that the greater the physiological compatibility of medicinal agents the greater their remedial efficiency.

But, as just intimated, this vital compatibility is of a compound character, being both of a material and dynamic nature, and manifested either singly or in conjunction with each other. Thus, for instance, such substances as iron, lime, potash, and soda, have an immediate material connection with the animal organism; while, on the other hand, quinia, strychnia, and others of the same class have a somewhat direct dynamic relation thereto; light, heat, electricity, and all influences of a psychical nature being still more purely dynamic in their properties and effects; whereas phosphorus and some other agents of a similar character, subserve both a material and dynamic purpose in the processes of life.

With a view, therefore, to concentrate attention upon a remarkable agent of the kind last indicated, as well as to intensify thought upon the general subject of the physiological compatibility of remedies, I propose in this paper to present some general observations upon the medical properties and applications of nitrous oxide,

---

* Many of them in fact, acting thus in the double capacity of materia alimentaria and materia medica.

referring those interested for a more extended notice thereof to my former publications respectively entitled Zoo-adynamia; Toxicological, but which should have been Antidotal Applications of Nitrous Oxide,* *Boston Med. and Surg. Jour.*, vol. xlvi. No. 14; Anæmatosis, its consequences, prevention, and treatment, Ibid., vol. xlvi. Nos. 22, 23; Experimental Investigations on the Antidotal and Revivifying Properties of Nitrous Oxide, Ibid., xlvii. No. 19; Hæmatosis, its natural and artificial induction, Ibid., xlix. Nos. 3, 4, 5, 6; Glucosis, Ibid., vol. L. No. 11; Nitrous Oxide, its properties and applications, *Dental Cosmos*, vol. i. No. 12; Nitrous Oxide, its medical properties and applications, *Boston Med. and Surg. Jour.*, vol. lxvii. No. 25; and *Amer. Med. Times*, vol. vi. No. 6; Nitrous Oxide in Asphyxia, *Med. and Surg. Reporter*, vol. ix. Nos. 21, 22; Nitrous Oxide as an Anæsthetic, *Dental Cosmos*, vol. v. No. 5.

In the effort, therefore, to render this exposition as succinct, comprehensive, and practical as possible, I will treat of the subject under the several heads, of first, the chemical constitution, properties, and correlations of protoxide of nitrogen; second, its physiological influences and hygienic uses; third, its medicinal properties and applications, therapeutic, revivifying, antidotal, and anæsthetic; fourth, its preparation and combinations; and, fifth, its modes of administration and dose.

---

* Several typographical errors occur in this and other papers, partly from not having had an opportunity to see the proof.

I. Chemical Constitution, Properties, and Correlations of
Nitrous Oxide.

In the first place, with regard to its constitution,
nitrous oxide is a chemical compound in equivalent
proportions of the two gaseous elements nitrogen and
oxygen, hence designated in accordance with the usual
nomenclature and notation protoxide of nitrogen, with
the symbol NO, and equivalent numbers 22,00.   It is
an elastic, colorless gas of the sp. gr. 1,527, having a
somewhat faint, but agreeable odor, and sweetish taste,
which it imparts to water.   Under a pressure of about
30 atmospheres at 0°, or 50 atmospheres at a temperature of 45° F., this gas condenses into a colorless
transparent liquid, and from between 100° to 150°
below zero solidifies into a beautiful clear crystalline
body.   The evaporation of this solid protoxide of nitrogen produces a degree of cold far greater than that
of carbonic acid in vacuo, yet as it evaporates slowly
it does not like the latter, solidify by its own vaporiza-
tion.*

---

* Some recent observations upon this subject in confirmation
of the above, are thus detailed in a late number of the *Med.
Times and Gazette,* from the *Rev. Med.*

"*Liquefaction of Laughing Gas.*--One of the most interesting
objects at a recent *soirée* at the Paris Observatory consisted in
the exhibition of the liquefaction of laughing gas, the protoxide
of nitrogen, by M. Bianchi.   This took place at zero Centigrade
under a pressure of thirty atmospheres, the fluid issuing in a

The fact that protoxide of nitrogen is capable of being thus somewhat readily reduced to a liquid and solid state is now of very little more than scientific interest; but the time is probably not very far distant when it will become of the utmost practical value in view of the many and highly important medical purposes to which this remarkable agent is applicable.

In composition nitrous oxide differs from all other chemical bodies, although identical in constitution in the main, with atmospheric air, varying therefrom, however, both in the proportion of its constituent elements, and in the character of their association. Thus while nitrous oxide contains about one-third of oxygen to two-thirds of nitrogen, atmospheric air has only about one-fifth of the former to four-fifths of the latter. Moreover, in nitrous oxide the respective elements nitrogen and oxygen, are in chemical combination with

---

small jet from a strong metallic reservoir. Received in a glass tube, it retained its liquid condition by reason of the depression of temperature produced by evaporation, so that mercury being introduced solidified, and could be hammered like lead. Simultaneously, a body in a state of ignition, plunged into the atmosphere of the liquid, in which the mercury froze, burnt with a brilliant light. On pouring the protoxide into a small platinum capsule heated to redness, the liquid was found to retain all its properties while assuming the spheroidal state, and was still able to freeze mercury contained in little glass ampullæ. Finally, the liquid protoxide became solidified under the recipient of an air-pump, the temperature being reduced to 120° below zero Centigrade—the most intense cold yet obtained.";

each other, whereas in atmospheric air they are in but simple mechanical association without any apparent chemical union whatever. Nevertheless, though thus differing in the relative proportion and character of association of these constitutional elements, protoxide of nitrogen and atmospheric air are similar in their general properties and relations, varying more in the degree, perhaps, than in the nature of their affinitive reactions and physiological effects.

Besides atmospheric air, nitrous oxide is closely correlated with oxygen, to which indeed it is so directly identified as to encourage the belief that all its active properties as well as those of its congener—atmospheric air—depend exclusively upon this one element, but that such is not the fact we expect to make evident hereafter, notwithstanding the apparent similarity throughout of these respective agents, for in most of their prominent features they resemble each other so strikingly as to give rise to the impression of their interdependence upon one and the same substance just mentioned. This general similitude is especially manifest in the chemical and vital reactions, for like atmospheric air and oxygen, protoxide of nitrogen is an active supporter of combustion and of life, though in these respects it is in some measure more nearly allied to the latter than the former, from the greater relative proportion of this important element as well as in consequence, doubtless, of the peculiar combination of its constituents.

While, however, there is thus an intimate mechanical, chemical, and physiological correlation between these respective gaseous bodies, there are some specific differences in nature and properties manifested more particularly in their vital influences, which render them appropriate for distinct though somewhat similar medical purposes. But as a notice of these more in detail involves the consideration of another branch of the subject, I will proceed to their further discussion in that connection.

II. Physiological Influences and Hygienic Uses of Protoxide of Nitrogen.

1. The *physiological influences* of nitrous oxide are of a peculiar and striking character, for though like some other agents it exerts a very energetic and decidedly stimulant action upon the animal economy, yet this is in general so entirely distinct from all other excitants as to be quite unique.

The effects of protoxide of nitrogen upon the human system vary in proportion to the quantity appropriated and the particular susceptibilities or conditions of individual organisms, passing from a gentle acceleration of all the functions of the body to a high degree of physical excitement and mental exhilaration amounting in the extreme to an intensely pleasurable delirium or ecstacy which may indeed become so pure and exquisite as to absorb the consciousness of existence itself.

When exhibited in moderate quantities, nitrous oxide usually produces a mild and very pleasant thrilling sensation rapidly extending over the whole system, attended with forcible and prolonged respiratory efforts, and a strong disposition to laughter and muscular motion, which often becomes so irresistible as to be involuntary. If taken freely, these agreeable sensations are greatly enhanced, accompanied with a remarkable buoyancy of spirits, boisterous gayety, activity of imagination, rapid flow of vivid ideas and brilliancy of mental conceptions of the most sublime character, the direction of thought being influenced in some measure, by the dominant idea at the commencement of its action, hence sometimes followed by other than hilarious manifestations. Used largely, this state of excitement rapidly passes into one of rapturous enjoyment, entranced tranquillity, and complete repose with entire insensibility and unconsciousness, during the existence of which the mind becomes temporarily oblivious to all impressions, even of an ordinarily painful nature. The primary stage of sur-excitation with its concomitants, anæsthesia and trance, is usually, however, of brief duration, terminating rather suddenly, yet leaving generally a sense of permanent invigoration similar to that resulting from a free exposure to fresh atmospheric air, not being followed with any reactive languor or depression so common with ordinary stimulants.

Besides its general physiological effects, nitrous oxide

has also special tendencies to certain parts of the body, particularly the blood, brain, nervous system, and genito-urinary organs, being very efficient in preserving the healthy integrity of such portions of the economy and in promoting the functions more immediately connected therewith.

Respecting the *modus operandi* of protoxide of nitrogen, there is very little doubt but what it exerts both a material and dynamic influence upon the animal organism through each of its constituents singly and conjointly, yet, notwithstanding the peculiar character of its biological effects seems to conclusively prove that they are dependent upon its constitutional elements in their separate as well as combined state, it is, by some, thought to act through one exclusively. Nevertheless, though it is obvious that much of its potency is derived from the oxygen, it is demonstrable that it alone is not sufficiently energetic to account for all of the phenomena resulting from the operation of nitrous oxide upon the human system, some of these being so entirely distinct from the usual action of that element as to justify the conclusion of the additional influence of the nitrogen for their production.

This is especially the case in its marked tendency to, and specialty of action on particular parts of the system, as for instance the genito-urinary organs, wherein it differs greatly from the former, for it not only increases the excitability of these organs, but improves the quality as well as regulates the quantity of secretion therefrom, both by directly augmenting their power and promoting the normal genesis and elimination of their educts. The correctness of this view is more clearly apparent from the peculiarity of the effects of nitrous oxide upon the renal secretion in the enlarged production of the nitrogenous ingredients—especially urea—of that fluid, to which it gives rise in a remarkable degree. It is most probable, therefore, that the physiological action of protoxide of nitrogen is of a compound character, from the operation of both its constituents—nitrogen and oxygen—in their associated and single state, the affinitive reactions between them and the general components of the organism being more decidedly manifested in consequence of its undergoing decomposition, and the presentation of these elements in a nascent state within the body. The chemico-organic and bio-dynamic influence of nitrous oxide is hence so much the greater from the presence of its constituents in an actively divided state, the nascent condition affording the most favorable opportunity for chemical reaction, molecular nutrition, organic metamorphosis, dynamic manifestation, and vital development.

Protoxide of nitrogen is thus altogether unique in being a rapidly diffusible, potent, general, and permanent stimulant, intensifying all the vital actions, supplying elements of nutrition, exerting both a material and dynamic influence upon the animal economy, and in having a direct physiological compatibility therewith both functional and organic.

Hence, with respect to the predominant characteristics of protoxide of nitrogen it is truly *sui generis*, though closely allied in chemical constitution, material properties, and sanitive effects with atmospheric air, which may be regarded, in fact, as its natural prototype, differing therefrom apparently more in the proportion of its constituent elements—nitrogen and oxygen—and in the manner of their association, than in any other essential respect, although its action upon the vital economy is more energetic, definite, and perceptible, than that of the latter, varying rather in the degree, perhaps, than in the nature of its physiological influence. Still, though there is thus a very strong analogy between protoxide of nitrogen and atmospheric air, yet in their relative effects upon the animal organism they vary somewhat materially, as there is a striking difference in the ratio or measure as well as in the kind of action thereupon, for the biological effects of the former are not only more intense, concentrated, and manifest, but also more decidedly stimulant than those of the latter, though they are likewise of a general, very bland, and highly invigorating character.

While, therefore, there is thus a very intimate correlation between these two compound gaseous bodies—nitrous oxide and atmospheric air—yet each one has special peculiarities of its own, of such a marked character indeed, as to make them appear quite distinct and render them useful for diverse as well as similar purposes.

2. The *hygienic uses* of protoxide of nitrogen are varied and important. The nature of these is indicated by the character of its physiological influences in accelerating the normal processes of life, yet for practical purposes a more specific notice thereof may not be unprofitable.

In brief then, through its constituent elements and dynamic properties, nitrous oxide exerts a powerful influence in both supplying essential matter for organization, and in promoting the general molecular, cell, nutrient, reproductive and dynamic operations of the animal economy, those of the vegetal, animal, and psychical life inclusive. It is thus indeed, remarkably active and potent in promoting the various functions of digestion, absorption, circulation—both general and capillary—aeration or arterialization, hæmatosis, calorification, assimilation, disintegration, depuration, secretion, excretion, muscular and general contractility, innervation, and intellection; and likewise, those of the reproductive system. Hence, for the preservation of the healthy integrity of the body and the regulation as well as invigoration of all the important functions of life, this agent may, *cæteris paribus*, be always employed with advantage. This concise but comprehensive outline of the sanitive effects and applications of protoxide of nitrogen will suffice to show that in an hygienic point of view it is both unique and invaluable.

It would seem, *a priori*, that an agent which exerts such a potent material and dynamic influence over the animal organism in maintaining its normal status, should necessarily be very efficient in correcting as well as in preventing abnormal action, and both observation and experience have shown the correctness of this conclusion. It has thus been demonstrated, both *a priori* and *a posteriori*, that nitrous oxide is of pre-eminent importance in restoring as well as in preserving health.

Of these remedial powers and applications I will now speak more specifically.

III. Medicinal Properties and Applications of Protoxide of Nitrogen.

As this branch of the subject—which, like the preceding, involves principles and practice of a perfectly novel yet eminently practical character—is so very extensive as to require for its proper elucidation more time and space than can now be given it, I shall not attempt to enter much into detail, but merely endeavor to present a very general sketch of the prominent points of interest, leaving more specific observations for some future period. With a view, therefore, to thus exhibit a somewhat condensed and systematic summary of the medicinal relations of nitrous oxide, I will, as before mentioned, treat of it in the several aspects of its therapeutical, revivifying, antidotal, and anæsthetic influences and uses.

1. In a *therapeutic* as well as in an hygienic point of view, nitrous oxide is of extraordinary interest and value. Its practical applications for the removal as well as the prevention of disease are numerous and diversified. In fact, in these respects, the protoxide of nitrogen is not surpassed, if equalled by any known sanitive agent outside of those included in the materia alimentaria. It is indeed *sui generis*, as in consequence of its peculiar chemical constitution and properties, its specialty and potency of physiological action, its extensive range and variety of sanitive application — hygienic, therapeutic, revivifying, antidotal, and anæsthetic—it differs greatly from, and is superior to the best and most powerful of known remedies. Nitrous oxide is thus unique in its physiological and sanitive influences as well as in its chemical constitution and properties; for, though analogous in the extent and variety of its therapeutic uses, to some of the most active medicinal substances, such as iron, quassia, strychnia, quinia, mercury, etc., it is far superior to them in the greater range and diversity of remedial application as well as in the general character and special peculiarities of its effects upon the animal organism. Indeed, from my present knowledge upon

the subject I am convinced that protoxide of nitrogen will supersede, to a considerable extent, some of our most reliable and popular remedies, while at the same time it will render the prevention and resolution of many of the ordinary forms of disease more certain, speedy, and decided; and, moreover, afford the means of removing some of those peculiar abnormities now not at all or but slightly amenable to the present therapeutic measures.

In view, therefore, of its peculiar and valuable sanitive properties, nitrous oxide promises to be a very efficient general substitute for some of the most potent and expensive remedial agents known, such as for instance, alcohol, ammonia, quassia, strychnia, mercury, and others, variously classified as diffusible and permanent stimulants, tonics, antiperiodics, antispasmodics, alteratives, secernents, etc. Among those which it may thus more or less completely replace is that peculiar and valuable remedy *quinia*, and as a succedaneum, therefor, as well as in some measure the other medicinal agents referred to, I present it for particular consideration.

While, however, it will thus, to a great extent, become a substitute for, and better subserve the sanitive purposes of a number of potent remedies, it cannot, of course, meet efficiently the necessities of the case where there is a privation of the normal components of the economy otherwise than those of its own constituents, for in the absence of some simple or compound organic substance like lime, iron, phosphorus, albumen, fibrin, etc., it could not be expected to supply the deficiency, although even here it will often prove indirectly service-able by rousing the vital energies to that degree of activity necessary to their appropriation or develop-ment from the regular sources of nutrition, as it directly increases the strength of the body, stimulates the func-tions of assimilation, improves digestion, and sharpens the appetite, while at the same time furnishing within itself certain important elements of aliment, usually supplied in a gaseous state by the atmospheric air through the pulmonary organs; nutritive, like other matter being in the several forms of solids, liquids, and gases, principally introduced into the more complex organisms through the common channels of the stom-ach and lungs. Protoxide of nitrogen thus acts in the double capacity of a nutritive and remedial agent com-bining the properties of an article of both the materia alimentaria and materia medica, presented in a form suitable for hygienic and medicinal purposes. Hence presupposing the due proportion in quality and quan-tity of other essential elements of alimentation, nitrous oxide will always, *cæteris paribus*, prove more or less useful within the range of its capabilities both mate-rially and dynamically.

But passing from such general reflections to a par-ticular consideration of the medicinal influences of this

remarkable agent, I will, in order to give a more definite view of the special properties and applications of
protoxide of nitrogen, present a general outline thereof
so far as the current imperfect nomenclature will per-
mit. Thus in comprehensive terms, nitrous oxide is a
direct, potent, and permanent chemico-organic, arterial,
nervous, cerebral, and general stimulant, secernent, depurant, aphrodisiac, and antitoxic, having a special
tendency to the blood, brain, nervous system, and genito-
urinary organs. It exerts a powerful invigorating influence over the entire economy and is a superior nutrient, hæmatic, neurotic, tonic, disintegrant, diuretic,
disinfectant, alterative, resolvent, sorbefacient, antidote,
antiseptic, etc. etc. Its primary action is usually prompt
and frequently well marked, though somewhat transi-
tory in character, while its secondary or more remote
effects are permanent and highly salutary, the difference
between them being more in degree than in kind, for,
as before stated, the invigoration is generally con-
tinuous and persistent without subsequent depression,
as is the case with most other stimulants. The properties and influences of nitrous oxide are, in other words,
both organic and dynamic; organic in supplying the
material elements—nitrogen and oxygen—for the various chemical purposes of the economy; and, dynamic
in stimulating the functions of the whole body. Thus
both through its constituent elements and dynamic influences it promotes the chemico-organic and dynamic

operations of the animal organism, and while thereby
regulating normal, resolves as well as prevents abnormal action.

In consequence, therefore, of its peculiar chemical
constitution and sanitive properties, protoxide of nitrogen is especially applicable to the correction as well as
prevention of numerous derangements of an atrophic
and adynamic character, whether primary or secondary,
antecedent or consequent. Its power of thus averting
and removing disease of an asthenic nature more particularly, both general and local, organic and dynamic,
primary and secondary, acute and chronic, is—according to my own experience—certainly very great and
often strikingly manifested. From its marked practical
value in the prophylactive and curative treatment of
the more common types of disease it will, doubtless,
also prove extremely useful in preventing and resolving
some of those of a peculiar character now not at all or
but partially, amenable to the ordinary remedial influences.

In general nitrous oxide is of great utility in the
treatment of those asthenic conditions in which the
material and dynamical processes of the animal economy are in abeyance, and which are so frequently exhibited in the inertia of the various functions of the
organism, those of the vegetal, animal, and psychical
life inclusive. It is therefore especially indicated in
indigestion and inefficient absorption, as also in general

inactivity of the chylopoietic functions; in imperfect aeration or arterialization of blood and deficient hæmatosis; in mal-assimilation and disintegration; in insufficient secretion and depuration; and, in irregular or defective motility, contractility, innervation, and cerebration. Hence in the various forms of asthenic dyspepsia and other morbid states dependent upon or associated with torpidity of the chylopoietic viscera; anæration, anæmatosis, and mal-nutrition generally, both primary and secondary; in depraved and defective secretion and depuration; in enervation, neuralgia, chorea, paralysis, melancholy, amentia, and adynamic states generally, the nitrous oxide will, doubtless, always prove more or less useful as a curative agent.

Protoxide of nitrogen is moreover, strongly indicated in atonic conditions of the genito-urinary apparatus, more especially in inertia and such other abnormities of the urinary and reproductive organs as are presented in cases of incontinence and suppression of urine, paralysis of bladder, spermatorrhœa, impotence, sterility, some forms of amenorrhœa, dysmenorrhœa, leucorrhœa, menorrhagia, etc., as it has a special tendency to these organs, and exerts a powerful influence over their functions.

But the applications of nitrous oxide are not exclusively limited to the more purely atrophic and adynamic states, as it is available for the successful treatment of some forms of both general and local hypertrophy, and

mal-organization. Thus, for instance, in the undue or
abnormal production of adipose, fibrous, and other tissues, as in obesity, enlargement and fatty degeneration
of the heart and other parts, it seems to act very efficiently by superoxidation or otherwise in resolving such
abnormities, and in restoring the equilibrium of nutrition, innervation, contractility, and tonicity. Moreover, in various analogous conditions, such as elephan-
tiasis, phlegmasia dolens, and similar general and local
hypertrophies, it will doubtless also prove useful as a
remedial if not a curative agent.

Among those general morbid states to the curative
treatment of which the nitrous oxide is more particularly applicable are those which precede and give rise
to, or are coincident with various constitutional disorders, such as scrofula, consumption, and other affections of a like character. It is not only useful in correcting such cachexia and in averting their sequelæ,
but likewise, to a considerable extent, in resolving these
latter. This agent, protoxide of nitrogen, will doubtless also remove as well as prevent other abnormities of
an asthenic character which are more or less dependent
upon a special diathesis, and besides favorably modify,
if not entirely subvert the latter. But as I have already referred to some of these and a special detail will
unduly extend this communication, I will merely allude
in a very general way to such as are most prominent,
and refer those interested for a more extended notice

thereof to the published papers before mentioned, particularly to those respectively entitled Anæmatosis, Hæmatosis, and Glucosis. Thus, for instance, in those forms of deranged nutrition and innervation attended with an abnormal production of fibrin and fat, as in polysarcia, and especially in that variety known as adiposis, the nitrous oxide is of great practical value in causing these substances to undergo their ultimate metamorphosis and final disorganization, and in restoring the healthy balance of the economy. Through its chemico-organic and bio-dynamic action protoxide of nitrogen will likewise transform and disintegrate, as well as prevent the undue production of albumen and sugar, and thus counteract their attending diseases, albuminuria and glucosuria with their sequelæ, the development of all of which I regard as dependent in the main, upon a constitutional dyscrasia, believing that a similar diathesis is generally concerned in the excessive production of such normal substances as sugar, fat, albumen, fibrin, etc., as in that of an abnormal matter like tubercle or cancer, although they might also be engendered independent of any direct constitutional predisposition.*

But in the treatment of the special manifestations of such cachexia—which are numerous, diversified, and complicated—nitrous oxide is also of great practical

---

* This view, which was formally presented years ago, is supported by the results of recent investigations.

value, for though not altogether adapted to every degree and variety of these secondary affections it is of great general utility in many if not all. Hence to derive the utmost benefit from its use in such cases it must be exhibited with due regard to the particular type, stage, intensity, and complications of the local disorder, for notwithstanding always more or less strongly indicated in these maladies to correct the constitutional derangement and subvert, remove, or at least modify its destructive sequelæ, yet it is not always admissible in every stage of the secondary lesion, especially when of an actively irritable or inflammatory nature and of a sthenic character, but in the subacute and chronic condition, particularly of an asthenic type, it is mostly of superior value.

Thus, for instance, in phthisis, in which it is very useful in promoting healthy arterialization, hæmatosis, assimilation, secretion, and innervation, and in relieving oppression of breathing, cough, and other distressing symptoms, it must be exhibited with due regard to the local complication, for if given too freely during active inflammation or tuberculous colliquation, it may promote the destructive process and hasten the fatal termination, whereas if employed in proper quantities with just discrimination, it will tend in all stages, to diminish morbid action, remove abnormal matter, rectify constitutional derangement, and restore the normal organic and dynamic status of the economy.

Again in another class of secondary affections connected with the abnormal deposit or undue production
of such normal substances as sugar, fat, albumen, etc.,
in which it is very efficient, nitrous oxide must be employed with the same restrictions in order to derive the
utmost benefit. This is particularly the case in those
disorders engendered by a more immediate eliminative
effort on the part of certain organs like the kidneys, to
remove from the body an excess of such matters, and
which are manifested in the various forms of glucosuria,
albuminuria, renal irritation, congestion, inflammation,
degeneration, and disorganization with their destructive concomitants. In these cases in suitable quantities, properly exhibited, nitrous oxide will doubtless
always prove more or less useful not only by directly
transforming, disintegrating, and preventing the further
excessive development of such offending substances, rec-
tifying the constitutional cachexia, restoring the healthy
balance of innervation, nutrition, secretion, and elimination, but also by resolving more or less completely
the consequent local lesion.

As a curative agent in glucosuria · of a mild form
nitrous oxide is especially active, for I have repeatedly
seen the saccharine condition of the urine rapidly dis-
appear under its use, but of its remedial efficiency in
the extreme cases of this affection, I am unable to
speak from practical experience, never having had an
opportunity of treating such an abnormity. Neverthe-

less, for the cure as well as prevention of all forms of
that peculiar malady known as diabetes—except perhaps, the traumatic variety so long as it is thus directly
dependent upon the state of traumatism, or in cases
of great disorganization of structure—I am confident
protoxide of nitrogen will, *cæteris paribus*, prove a
specific. Moreover, I firmly believe this agent will
also materially modify if not entirely remove the various tuberculous, cataractous, and other sequelæ of the
glucosic cachexia. Hence, with the exception mentioned, in every grade and variety of the glycogenic
disorder, even if attended with extreme modification of
structure, nitrous oxide promises to prove of great
remedial value. For the curative treatment, therefore,
of the glycogenic diathesis and the various derangements of the body connected therewith, both functional
and organic, and especially those forms known as glucosuria, diabetic cataract, and other concomitant local
lesions, protoxide of nitrogen is particularly recom-
mended as a very promising remedy. It may probably
also exert a beneficial influence to some extent, at least,
in modifying or resolving other varieties of cataract
and analogous alterations of structure in different parts
of the economy.

In the numerous sequelæ of such diatheses as scrofu-
losis, adiposis, albuminosis, and others of a like character so frequently presented in the forms of depraved
nutrition, defective or degenerate structure, modified

secretion, and the various inflammatory complications of the different tissues and organs, nitrous oxide is also generally applicable with the same restrictions as to the special contraindicating circumstances before mentioned. Hence, *cæteris paribus*, in scrofula, fatty, waxy, and similar modifications of tissue, albuminuria, chyluria, oxaluria, etc., with or without dropsy, it may be exhibited somewhat freely, but in the more acute and advanced states of sthenic irritation or inflammation it should be carefully employed, if at all, in very small quantities, for though always more or less strongly indicated to correct the constitutional dyscrasia, promote normal metamorphosis, resolve local lesions, restore and regulate the healthy action of both the general system and particular part affected, yet if given unduly or inappropriately it will be apt to increase the inflammatory action and the tendency to disorganization of the implicated structure. The use of protoxide of nitrogen is therefore mostly contraindicated in all such actively irritable or inflammatory states, even of an asthenic nature, and especially of the cerebral, cardiac, hepatic, or renal organs. This is particularly the case in the several nephritic complications of the character referred to, and among others the so-called Bright's disease in which nitrous oxide may frequently be employed with benefit, if given in small quantities at a time and with such other precautions as not to increase uræmia or unduly stimulate the affected organ and general system.

The same principles apply with even greater force in another species of secondary disorder of which rheumatism and gout are examples. In such abnormities nitrous oxide is also highly useful in correcting both the constitutional and local derangement when of a chronic character, but is as a rule, inadmissible in the acute form in consequence of the active tendency to cardiac inflammation, and for the same reason more or less objectionable in the subacute condition, though to a certain extent beneficial therein, when exhibited in moderation and with due discrimination. In the first variety, however, notwithstanding it occasionally seems to temporarily develop or augment pain* and excitement from a disproportionate quantity, perhaps, yet by modifying the general dyscrasia, removing the immediate cause of disturbance, and restoring the healthy equilibrium of system, it sometimes speedily resolves the abnormity altogether.

It will thus be seen that protoxide of nitrogen has a very wide range of therapeutical application in that class of maladies connected with the undue or irregular production of such substances indicated, especially when of an amyloid, glucoid, ceroid, adipoid, albuminoid, or

---

* That it will thus sometimes momentarily develop pain was long since observed by Sir Humphrey Davy, *vide* " Researches on Nitrous Oxide."

pigmentary character. Hence in the various concomitant lesions of nutrition and innervation manifested in the different modifications and degenerations of structure from abnormal formative or disintegrative metamorphosis, whether of a benign or specific nature, and particularly in the many analogous abnormities of the kind in the tissues of the hepatic, cardiac, ophthalmic, renal and other organs and parts of the body, nitrous oxide will, no doubt, prove as generally useful as it has heretofore specially.

As already intimated in those other diathetic derangements of nutrition and secretion attended with the undue or extraneous development of such substances as cholesterine, oxalic acid and their analogues, nitrous oxide will doubtless also prove practically useful in correcting the abnormal and in restoring the normal action of the molecular and somatic life as well as the special functions and organs more immediately implicated. From its superior power in modifying general nutritive, dynamic, and secretory action, I believe protoxide of nitrogen will thus prove of great value in preventing and, perhaps, also in causing the destructive disintegration as well as expulsion of certain adventitious concretions in different parts of the body, and particularly some of the varieties of biliary and urinary calculi. The special type or variety of lithiasis to the treatment of which it seems most applicable is that resulting from an organic deficiency of its constituent elements, oxygen

and nitrogen, although it is not improbable that it may
likewise be efficient in preventing the development of
other forms of calculi in which these elements are not
so immediately concerned through its general chemico-
organic and bio-dynamic influence in modifying morbid
constitutional and local tendencies, and in promoting
healthy metamorphosis, innervation, depuration, and
elimination.   Hence, as an antilithic it may be resorted
to in the confident hope that it will often meet to some
extent, at least, both the general and special indications and be thus directly and indirectly useful.

That protoxide of nitrogen will act very promptly
and efficiently in promoting hæmatosis and the
general assimilative, disintegrative, and excretory
processes—particularly those more immediately connected with the renal organs—I am well satisfied,
from much observation and experience.   Indeed, in
these respects it is not surpassed, if equalled, by any
other known agent, and especially in the character
of its effects upon the kidneys, as it not only actively
promotes the secretion of urine, but also greatly influences its healthfulness, both in quantity and quality.
Nitrous oxide is thus, in fact, a most potent and
peculiar diuretic ; for, unlike other remedies of this
class, while it directly increases the solid components
of the urine, and particularly urea, it at the same
time regulates and facilitates the elimination of the renal
secretion, arterializes and purifies the blood, gives tone

to the tissues, strengthens the nervous, and invigorates
the whole system.   It is hence applicable to the pre-
ventive and curative treatment of quite a number of
urinary affections, — such, for instance, as anuria,
anazoturia, chyluria, oxaluria, and other varieties of
lithuria, dysuria, hydruria, hæmaturia, enuresis, cystor-
rhœa, paralysis of bladder, and others of a similar
character.

Besides those thus indicated, there are various other
abnormities, apparently unconnected with any special
diathesis, constantly presented in the different forms of
asthenic hyperæmia, congestions, inflammations, serous,
hæmorrhagic and plasmatic effusions, and depraved,
diminished, or excessive secretion within or from the
several parts of the body,—particularly of the pulmonary, alimentary, and genito-urinary organs, in which
nitrous oxide is also generally useful.   Hence, in the
several maladies of the kind, and especially in such
conditions as diffused anasarca and localized dropsy,
catarrhus, diaphoresis, diuresis, leucorrhœa, diarrhœa,
cholera, and similar internal and external defluxions, it
may frequently be employed with decided advantage.

This agent—protoxide of nitrogen—is also useful
for the removal of those constitutional disorders upon
which such affections are so frequently dependent,—as,
for instance, anæmia, hydræmia, chlorosis, and other
forms of mal-nutrition and debility.

While, however, the therapeutical applications of

nitrous oxide are so varied and important in this direction, they are not thus exclusively limited to the more purely vegetal or organic, as it is also of extensive use in the animal and psychical life,—being exceedingly valuable in adynamia of the brain and nervous system. Thus in the numerous asthenic derangements of innervation and cerebration manifested in the different forms and degrees of enervation, anæsthesia, neuralgia, imbecility, mental depression, hypochondria, delirium and similar mental states, it is generally applicable. By regulating vital action and intensifying all the functions of life, protoxide of nitrogen often proves very efficient as an anodyne, hypnotic, and general nervine, in correcting undue excitability of body and mind, relieving pain and suffering, promoting sleep, removing vital inertia and atony, and in restoring the healthy balance of the economy.

In addition to these aberrations of general innervation and intellection, there are others connected with special sensation, contractility, and motion, in which nitrous oxide is also more or less strongly indicated. It is hence applicable in spasm, torpidity, debility, and paralysis, whether local or general, partial or complete, acute or chronic,—as in chorea, epilepsy, amaurosis, hemiplegia, paraplegia, and analogous convulsive and atonic disorders. As an antispasmodic, antiparalytic, and corroborant, it is thus of general application in the treatment of all such affections dependent upon an adynamic state, either of molecular or somatic life, and not on defective nutrition, from the absence of some organic element, especially of the neural matter,—as phosphorus, fat, albumen, osmazome, etc.; for it is obvious that in all cases of starvation of brain, nerve, muscular or other tissue, neither protoxide of nitrogen nor any other agent can be of more than incidental service, unless they supply the very elements of structure required. Hence, in order to derive benefit from this or any other remedy, it is necessary to previously insure healthy alimentation by introducing into the economy the requisite elements of nutrition in such quantities and form best adapted to supply the special organic deficiencies; and secondarily, to resort to such extraneous measures as may be essential to promote assimilation and normal life action.

These very general remarks will serve to show that protoxide of nitrogen has an extensive range of therapeutic as well as hygienic application, and especially in subacute and chronic conditions of an asthenic nature. While, however, it is usually most applicable in such morbid states, yet it is not thus exclusively limited, as it is also serviceable in the treatment of many other disorders of an adynamic character, in their more active or acute, as well as in their secondary or atonic stage. But as the notice of this class of maladies involves to some extent, the consideration of the revivifying and antidotal properties and applications of nitrous oxide, I will proceed to point out its merits in this direction.

2. The *revivifying* properties of protoxide of nitrogen are also very peculiar and striking, though closely resembling those of its congener—atmospheric air. The special character of these might, indeed, be readily anticipated from its kown chemical constitution and chemico-organic and bio-dynamic influences; but, independent of all such *a priori* considerations, experience proves that nitrous oxide is, in fact, of great practical value in the treatment of many depressed, asphyxiated, and toxic conditions of the system, by supplying to the suffering organism those essential elements of alimentation,—oxygen and nitrogen,—restoring the healthy equilibrium of arterialization and innervation, promoting normal nutrition and depuration, and in accelerating the general organic and dynamic processes of life.

Protoxide of nitrogen is thus of great utility in relieving those abnormal states resulting from a privation or contamination of atmospheric air. Hence it is especially indicated in the numerous disorders engendered by impure air, as well as in those caused by a more or less constant deficiency of atmospheric air from defective ventilation, sedentary habits, and other unnatural modes of living. Besides these slow and insidious, though no less certainly destructive methods of strangulation, nitrous oxide will prove useful in those more rapid and extreme cases of asphyxiation consequent upon apnœa, hanging, and other forms of suffocation,—including, doubtless, that from drowning, although my own experiments in this direction have not been very satisfactory.*

This agent may also be employed with advantage in asthma, cyanosis, the asphyxia of newly born infants, and in suspended animation generally, whether partial or complete, temporary or prolonged, for so long as the blood remains uncoagulated and molecular life is sufficiently active to admit of revivification, even when so much in abeyance as to appear entirely beyond the reach of any stimulus, protoxide of nitrogen will afford a most efficient means of resuscitation, as it exerts a powerful influence in promoting all the functions of the animal economy,—especially those of arterialization, hæmatosis, circulation, respiration, innervation, and muscular and general tonicity.†

As, however, much of the danger in such asphyxiated conditions is dependent upon the generation and retention of noxious matters within the economy, nitrous oxide will not only be indirectly useful in subverting their effects, by directly arterializing the blood, stimulating the nervous and invigorating the general system, but also, more immediately active as an antidote, in chemically combining with or decomposing such delete-

---

* Boston Med. and Surg. Jour., vol. xlvii. No. 19.
† Boston Med. and Surg. Jour., and Med. and Surg. Rep.

rious substances, and in promoting their elimination
from the body.

3. The *antidotal* influence of protoxide of nitrogen
may, therefore, be said to be of a compound character,—
acting in the threefold capacity of neutralization, decomposition, and elimination.

The peculiar constitution and properties of nitrous
oxide thus render it not only powerfully antagonistic to
all enfeebled, depressed, and asphyxiated conditions,
but also more or less so to a variety of pernicious influences and numerous toxic states from divers noxious
agents, which may be generated within or be introduced
from without the body. The beneficial applications of
protoxide of nitrogen is hence not restricted exclusively to cases of suspended animation from privation
of atmospheric air and the retention of excrementitious
matters within the economy, but is likewise extended
to those adynamic and toxic conditions of system resulting from a poisoned-state of the fluids and solids of the
body and prostration of the vital energies from the
pernicious operation of certain deleterious agencies of an
internal or extraneous origin, whether of a so-called
malarial or miasmatic nature, and of an infectious or
contagious character. It is, therefore, particularly
indicated in the various ataxic and adynamic fevers,
such as the typhus, typhoid or enteric, congestive,
yellow, remittent, intermittent, and all others of a similar type. Nitrous oxide is moreover of general

application and quite efficient in scarlet fever, measles, diphtheria, variola, constitutional syphilis, erysipelas, gangrene, and kindred maladies. It may likewise prove more or less useful in sunstroke, pyæmia, purulent infection, puerperal fever, and necræmic and toxicæmic affections generally, those from septic poisons and the virus of venomous and rabid animals inclusive.

The stimulant, depurant, and antidotal properties of protoxide of nitrogen also render it peculiarly valuable in the treatment of another variety of toxicosis from the inordinate and continued use of alcohol, tobacco, and opium, as well as in those sudden and dangerous states of intoxication from over-doses of the same, and the poisonous effects of belladonna, aconite, hydrocyanic acid, chloroform, carburetted hydrogen, carbonic acid, and perhaps also all others of a like character.

From this cursory sketch it will be apparent that protoxide of nitrogen has a very extensive range of medicinal application, and, *cæteris paribus*, is well adapted to the curative treatment of all depressed, morbid, and toxic states in which a chemico-organic, arterial, nervous, cerebral, and general stimulant, nutrient, alterative, resolvent, absorbent, secernent, antiseptic, antitoxic, and revivifying influence is required.

4. As an *anæsthetic*, protoxide of nitrogen is also unique, differing essentially from all other agents of the kind in chemical constitution, physical properties, and physiological influences, for these latter are not only

chemically dissimilar, but always more or less directly
sedative in their action upon the animal organism;
whereas, the former is *ab initio* primarily and permanently stimulant, not even being followed, unless in
exceptional cases, with any of that languor or depression so peculiar to the others. Thus, for instance, in
regard to constitution, while most other anæsthetics are
composed principally of the elements hydrogen and
carbon, the sole and exclusive constituents of nitrous
oxide are oxygen and nitrogen. Besides in their relative
effects upon the living economy there is as great a
disparity between protoxide of nitrogen and all other
agents employed for the production of insensibility as
there is in composition. This is especially manifest in
the action of those representative anæsthetics,—chloroform and ether, for the former, as elsewhere stated,* not
only directly prevents aeration of the blood, but, doubtless, also deoxidizes that fluid and diminishes general
chemico-organic action, stupefies the brain, depresses
the nervous system, causes relaxation of the muscular
and other tissues, paralyzes the heart, and thus produces
death,—its tendency being, in fact, to prostrate the vital
energies and destroy life by direct and positive sedation;
while ether, though primarily somewhat stimulant to the
brain and nervous system, and less immediately active
in arresting oxidation and metamorphosis, and in in-

---

* Dental Cosmos.

ducing insensibility, stupor, and paralysis, is yet ulti-
mately depressing and destructive, much in the same
way as its congener. Hence the process of anæsthetiza-
tion as thus accomplished, is a process of devitalization,
and the anæsthetic condition a state of suspended
animation artificially produced, the truth of which is
demonstrated by the fact that this, approximate or
partial, often proceeds, notwithstanding the utmost
care, to complete and absolute death. These agents
are, therefore, positive and powerful sedatives if not
altogether in their immediate, certainly in their more
ultimate effects.

The physiological influence of nitrous oxide is, however, the reverse of this, for, instead of retarding, it on
the contrary increases oxidation of the fluids and
solids of the body, stimulates the brain and nervous
system, augments general and special sensibility, excites
muscular and general contractility, accelerates molecular
metamorphosis, promotes general nutritive and vital
action, invigorates the whole system, and acts as a true
tonic. Its effects in these respects are indeed so well
marked as to place it in direct antagonism to the
various sedatives and render it very efficient in not
only counteracting their depressing influence by direct
stimulation, but also to some extent as an antidote
thereto. There is, therefore, a wide difference both in
constitution and properties between these respective

4*

agents and nitrous oxide; for, while the former arrests, the latter promotes and intensifies life action.

But it may be asked, if such is the case, how comes it that protoxide of nitrogen acts as an anæsthetic at all? for it is well established that it will produce insensibility, a fact long since observed by Sir HUM-PHREY DAVY,* and abundantly verified by recent experimentation. This effect of nitrous oxide was indeed so apparent to DAVY, that it led him to suggest its applicability for the relief of pain from surgical operations, as the following extract from his work will show.† "As nitrous oxide in its extensive operation appears capable of destroying physical pain, it may probably be used with advantage during surgical operations in which no great effusion of blood takes place."

Notwithstanding, however, this recognition of the property of nitrous oxide to cause insensibility and his suggestion for its employment to allay pain, it does not appear that DAVY, himself, either made, or urged others to make, any practical application of his thought, and much less foresaw its extreme usefulness in this direction; nor even, that he sufficiently appreciated the importance of the fact to induce him to further examine the subject. With DAVY, therefore, this was apparently but a mere casual observation and sugges-

---

* Vide Researches on Nitrous Oxide.
† Ibid., p. 329.

tion without special investigation or apprehension of
the anæsthetic properties of nitrous oxide or general
anæsthesia. But it was far otherwise with another to
whom must be accredited the wonderful discovery of
anæsthesia, both on the principle of *a priori* appreciation and *a posteriori* demonstration. This was Dr.
HORACE WELLS, whose mind from the first was deeply
impressed with the idea of the possibility of speedily
producing general insensibility in order to render
operations painless, and the belief that some means
existed whereby this might be promptly effected and
suffering thereby completely obviated. This conviction
led to research, and the moment the fact was presented
to him that nitrous oxide was capable of producing
insensibility it was seized upon with avidity, and the
practical application immediately made by an *experi-
mentum crucis* on his own person in having a large
molar tooth extracted while under its influence, which,
as he anticipated, was effected without pain, thus
proving conclusively the correctness of his preconcep-
tions respecting the feasibility of anæsthesia.

With regard, therefore, to the relative merits of these
pioneers of anæsthesia there can be little doubt, for
while Sir HUMPHREY DAVY is clearly entitled to the
credit of having first observed the anæsthetic properties
of nitrous oxide, and suggested its application for the
relief of pain in surgical operations, to Dr. HORACE
WELLS is unquestionably due the immortal honor of

having made the first practical demonstration of anæsthesia by means of this agent, its primary selection for
the purpose being apparently more the result of accident
than design, for it is very doubtful whether he had any
previous knowledge of DAVY's observation, or had ever
before even thought of nitrous oxide in connection with
anæsthesia. His claim rests, therefore, upon originality
of conception as well as priority of exposition and practical application of anæsthesia. This great discovery
was thus the direct product of preconceived thought
and intelligent research on the part of Dr. WELLS, to
whom the world is thereby so largely indebted that it
cannot, by any posthumous honors, do more than ac-
knowledge its, obligations for benefits conferred, yet
may, to some extent, manifest its gratitude by an ample
endowment for the support of his bereaved and indigent
family, which it is hoped will be speedily done.

But to return from this digressive though pertinent
inquiry respecting the discovery of anæsthesia to the
consideration of the *modus operandi* of nitrous oxide
in the production of insensibility. As before stated,
the ordinary effects of protoxide of nitrogen upon the
animal economy are actively and permanently stimulant,
accelerating all the vital operations by increasing
chemico-organic and bio-dynamic action. Thus while
chemically it rapidly arterializes the blood and promotes
elemental interchange, molecular activity and organic
metamorphosis, dynamically it ·stimulates the nervous

system, sensorium and general functions of life. But
when taken freely the physiological processes are ac-
celerated to such a degree as to temporarily overcome
systemic excitability and cause partial interruption of
vital activity from overstimulation, the stimulus over-
balancing excitability or the momentum being greater
than the velocity, on the same principle as without
exhaustion a horse may be "taken off its feet," and its
pace materially diminished by immoderate driving, the
impetus being greater than its capacity for speed. Besides this there may probably be such an abundance of
carbonic acid engendered by the energetic oxidation as
to check in some measure chemico-organic reaction and
dynamization, and thus induce vital inertia and insensibility. Dynamically the sensorial centres may also be
so greatly overexcited as to be unable to respond to
any additional stimulus, or their functions be partially
suspended by local hyperæmia from undue activity of
the organs concerned in innervation and circulation, as
well as from the excessive quantity of carbonic acid
from superoxidation. Moreover, some of the effects of
nitrous oxide may result from the compounds of nitrogen with hydrogen and carbon, as ammonia, cyanogen,
etc., they being of a mixed stimulant and tranquillizing
character. In these and perhaps also some other more
occult modes, sentient impressibility may be so materially
diminished by protoxide of nitrogen as to cause partial
or even entire suspension of sensibility and conscious-

ness. This power of nitrous oxide to produce anæsthesia by superoxidation, overstimulation, etc., is quite distinct from that of all other agents more especially of the hydro-carbonaceous variety, for they induce the anæsthetic condition by non-oxidation and deoxidation of system, and by directly checking chemico-organic reaction and annihilating sensibility and consciousness. The former, therefore, increases while the latter diminishes life action, with, in both instances, the same general result of insensibility of body and unconsciousness of mind, though relatively as different in character from each other as sleep is from stupor or satiety from starvation.

The anæsthetic effect of protoxide of nitrogen is therefore, the result of vital exaltation instead of depression, being similar in character to that state of impassibility to injury engendered by exquisite pleasure, moral exaltation, great mental preoccupation and excitement, or undue concentration of nervous energy upon any one part of the system with diminished sensibility elsewhere, and vigorous tonicity of body, examples of which are presented in numerous physiological and pathological conditions, as for instance, in the exaggerated feeling with concomitant systemic insensibility of high social, religious, or political enthusiasm, and other powerful emotional states variously exhibited in the indifference to pain during spiritual abstraction or inspiration, heroic passion, fanatical frenzy, ecstacy, and to a certain extent, in hysteria, insanity and analogous disorders, in all of which there is frequently more or less perfect analgesia without coincident nor seldom even any great degree of subsequent depression, but sometimes only a mere feeling of lassitude with an active tendency to renewed energy.

■ Thus ordinarily with the anæsthetic action of nitrous oxide, which instead of being attended with concomitant or reactive sedation is usually followed by increased vital power, neither its immediate nor ultimate effects commonly causing any temporary prostration or permanent debility, but generally on the contrary, invigoration of both body and mind.

Instead, therefore, of resembling, nitrous oxide differs essentially both in constitution and properties from the usual anæsthetics, and is, in fact, directly antagonistic and antidotal thereto, its anæsthetic effect being the result of a stimulant, and not of a sedative action.*

Protoxide of nitrogen is thus not only superior in chemical constitution and the nature of its primary effects upon the economy, but also, in leaving a permanent feeling of general invigoration instead of that sedation always in a greater or less degree attending the action of ordinary anæsthetics; and, moreover, in being in a large measure, unattended with that imme-

---

* *Vide* Dental Cosmos, vol. i. No. 12, and Boston Med. and Surg. Journ., vol. xlvii. No. 19.

diate or subsequent danger which render the latter, in the main, so objectionable.

By thus unduly intensifying the general functions of life, and especially those of the brain and nervous system, protoxide of nitrogen overcomes systemic excitability sufficiently to render the body insensible to, and the mind unconscious of impressions which would other-• wise occasion pain and suffering. Although, as before stated, this condition of body and mind is quite distinct from that caused by ether, chloroform, and similar agents, it is yet in a measure attended with the same general results in the production of a state of insensibility sufficient in degree to admit of certain surgical manipulations without the usual concomitant pain. This state of anæsthesia from nitrous oxide is usually, however, of very brief duration, terminating speedily and somewhat suddenly in from one to three minutes,* and is hence better adapted for slight and short, than for large and protracted operations.

With regard, therefore, to the relative rapidity, efficiency, permanency, eligibility, and safety of protoxide of nitrogen as an anæsthetic, it may be stated that although it will frequently produce insensibility in from one to two minutes, yet it is somewhat uncertain and

* Occasionally it may continue longer, and in one instance, according to Prof. Chesebrough (Dental Register, vol. xix. p. 183), the anæsthetic influence was kept up "from twenty to thirty minutes."

not very persistent, lasting only about the same length of time, unless it is continued by repeated inhalations, which may be effected but not with the same facility as with other agents. Hence notwithstanding anæsthesia may often be thus more promptly induced, it is not, in general, so certain, profound, or permanent as that from other anæsthetics, and consequently its range of application is necessarily more limited for anæsthetic purposes. Its eligibility is also much less, as by the present methods and gaseous form, it is much more difficult to procure and neither so convenient to handle or administer as other anæsthetics, although this difficulty may probably be partially, if not wholly, overcome by its preparation in a liquid state. In safety it appears to be greatly superior to other agents of the kind, for though not altogether devoid of danger as asserted by some, it is doubtless much less in degree and mostly of an entirely different character from that of ordinary anæsthetics. Besides it has the additional advantage of being exempt from the pungent odor, sickness of stomach, and other minor disagreeable concomitants, of ether, chloroform, and their analogues.

While, however, the physiological effects of protoxide of nitrogen are usually of a highly pleasurable and sanitive character, it cannot, nevertheless, be indiscriminately employed with safety, for the artificial excitement of system rapidly engendered by its free administration, may not only prove injurious by directly increasing the tendency to irritation, hæmorrhage, and inflammation in the parts subjected to surgical mutilation, but may also develop latent pathological tendencies of a different as well as of a like character in other parts of the body, in persons with certain abnormal predispositions, to such a degree indeed, as to seriously injure health, if not absolutely endanger life itself.

The precise character and particular manifestation of such tendencies will, of course, depend upon the special predisposition of the individual system acted upon, but they will necessarily be most likely to appear in certain definite parts of the body in accordance with the peculiarities of action of the disturbing agent — nitrous oxide having, as before stated, a marked preference for the blood, brain, nervous system, and genito-urinary organs.

These brief observations will suffice to show the general character of the dangers to be apprehended from the undue or injudicious administration of nitrous oxide, yet as they may not be sufficiently definite for practical purposes, I will present a more detailed notice of these extraneous tendencies.

Thus, for instance, the undue excitement occasioned by the free or inappropriate use of protoxide of nitrogen, may produce both primary and secondary irritation, congestion, serous or hæmorrhagic effusion, and inflammation in different parts of the body, and espe-

cially in the brain and kidneys. Besides, principally
by superoxidation and overstimulation, it may cause
excessive disintegration and undue waste as well as
abnormal excitement of the system, even to destructive softening of the brain, nervous tissue, and other
important structures. Furthermore, by unduly accelerating functional action it may give rise to rupture of
the heart and blood-vessels, or disruption and other
mechanical derangements of important parts of the
organism. Moreover, through its powerful aphrodisiac
effects it may intensify sexual desire to such a degree
as to cause unpleasant exposure or even serious trouble.
It is probable, also, that dangerous intoxication might
sometimes ensue from the chemical reactions of the
elements of protoxide of nitrogen with those of the
body and the consequent formation in excess of such
compounds as carbonic acid, cyanogen, ammonia, urea,
and other substances of the kind.

In view, therefore, of the highly important considerations for protecting the general health and insuring
the safety of those subjected to its anæsthetic influence,
as well as for the morality of the patient and reputation
of the operator, nitrous oxide should always be administered with due care and precaution.

Notwithstanding, however, these apparently serious
objections to the free use of protoxide of nitrogen
would seem to strongly militate against its general
employment for anæsthetic and other purposes, yet they

are materially diminished in force by the compensating
fact, that it has, in some measure, the ability to counteract such tendencies, whether antecedent or consequent, through its material and dynamic power to
aerate, depurate, and increase the plasticity of the
blood, encourage healing by first intention, regulate
innervation, circulation, nutrition, contractility, and
other essential functions; and, by its general systemic
invigoration and highly sanitive influence, to protect
the living organism against any temporary or permanent injury. While in this way some of the dangerous
tendencies of this agent are counteracted, it is, nevertheless, always necessary to bear in mind that such do
exist and cannot even be slightly disregarded without
peril, for evil may follow when least expected; hence
an enlightened judgment and a judicious discrimination
are constant prerequisites in the administration of
nitrous oxide, in order to form a correct estimate of
contraindicating circumstances, and guard against injurious results.

Respecting the alleged deaths from the effects of
protoxide of nitrogen taken for anæsthetic and other
purposes, there is much doubt, for none of the reported
cases appear to be sufficiently well marked to justify a
positive record against this agent. From a special
inquiry into the history and character of those noted,
Mr. G. Q. COLTON, of New York, has arrived at the
conclusion that in no single instance can the death be

fairly attributed to the influence of nitrous oxide.* But a critical examination of the cases presented seems to disprove in some measure, this opinion, for notwithstanding it was found in one case that chloroform instead of nitrous oxide had been employed, and, therefore, the latter had nothing to do with the death of the patient, in two other instances it was apparently somewhat at fault. Thus in the case of S. P. SEARS, of New York, who died about two hours after its administration, it is probable that it hastened death by promoting undue pulmonary congestion, if not, also, to some extent tubercular colliquation, the post-mortem disclosing the existence of intense hyperæmia and great disorganization of the lungs with constitutional degeneration, the patient having been for some time seriously affected with consumption.   The same principle applies in the case of Miss BELL, of St. Albans, Vt., who, it is stated by Dr. GILMAN, of that place, "inhaled a small dose of the gas for sport (not for anæsthetic purposes) with several others, on Friday afternoon, January 29th; came out of it as well as any one ever does; attended a party the same evening; as well to all appearance as ever; full of life and frolic; was taken sick the next day (Saturday) and died on the Wednesday following."†
This death is attributed to cerebro-spinal meningitis,

---

* Dental Cosmos, vol. v. p. 490.
† Ibid.

and perhaps, justly so, yet it is quite probable that the excitant influence of nitrous oxide was injurious in so far as it promoted the disorder of the brain and spinal marrow, while the fulness of "life and frolic" so freely manifested after its use was both an immediate effect of its stimulus and a precursor of the fatal affection, it being a well-known fact that increased functional activity is a primary and constant concomitant of inflammation.   Hence it is more than likely that in this case also nitrous oxide was accessory to the death, even if it did no more than encourage an inflammatory tendency, and thus partially act as an exciting cause of derangement, the predisposition previously existing, of course, to such a degree as to render the extraneous stimulus from this or any other agent absolutely dangerous.*

These cases serve, therefore, to exhibit the character of the danger to be apprehended from the inappropriate or inordinate use of protoxide of nitrogen, and if not entirely conclusive within themselves are still sufficiently impressive to justify the preceding precautionary remarks, which it may be stated, were, in the main, presented in the pages of the *Dental Cosmos* some time before they were reported.

---

* Notwithstanding, however, the apparent contraindications to the use of nitrous oxide in spotted fever, it may probably prove beneficial therein, more especially in the primary stage of depression to excite reaction, and during convalescence to promote recovery.

These precautionary remarks are not, however, made with any disposition to undervalue this remarkable agent or excite undue apprehension respecting the potency of its action upon the economy, but simply with a view to afford a correct exposition of its anæsthetic and other medical properties, for no one has a higher appreciation of its intrinsic merits and valuable sanitive effects than myself, yet it is proper that the truth should be known, to enable all to avoid the evil and obtain the good.

IV. Preparation and Combinations of Nitrous Oxide.

There are several methods of obtaining protoxide of nitrogen, which are well known to chemists, and described in the general works on chemistry, but perhaps the simplest and best plan is by the decomposition with heat of nitrate of ammonia. The necessary apparatus for its production and preservation consists of a generator, purifier, and receiver, which should be of glass, or other non-oxidizable substance, though the first may be of iron, the second of wood or some metal properly protected or capable of resisting the action of the agents employed, and the third of the same, or better still, of India-rubber or gutta-percha, where the gas is not required to be kept for any great length of time. The requisite heat may be obtained from various sources and be placed in direct or intermediate contact with the

retort. For the production of a moderate quantity of
nitrous oxide a sufficiency will be furnished by an ordinary alcohol or other lamp with one or more large
wicks, but for its free and rapid evolution it is better—
as in fact it is in all cases—to apply the heat through
the medium of a sand bath, which regulates and distri-
butes the temperature so equably as to diminish the
danger of injury to the apparatus or the production of
contaminating substances, for if too suddenly or intensely applied it is apt to fracture or explode the retort, unless composed of some strong material like
metal, as well as to cause the generation of noxious
matters.

Nitrate of ammonia is decomposed at a temperature
of between 400° and 500° F., but if the heat be too
great this salt will be volatilized and wasted, as will be
indicated by the appearance of a white cloud instead of
a colorless gas, or its decomposition will be attended by
the formation of such objectionable compounds as nitric
oxide and hyponitrous acid, either of which will, of
course, necessitate the reduction of heat and careful
purification. The retort should have a stopper, or be
so connected with the purifier as to be readily separated
therefrom, in order to speedily equalize the atmospheric
pressure when the heat is undue or is withdrawn, and
thus prevent destruction or flooding of the former by
the forcible reflow of liquid from the latter.

When pure, *cæteris paribus*, nitrate of ammonia is

resolved into protoxide of nitrogen and water, but as it
is sometimes adulterated, besides the deleterious agents
mentioned, other impurities may be given off, such as
nitric acid and chlorine. For ordinary purposes nitrous
oxide may be sufficiently freed from extraneous matter
by passing it through water, with a large quantity of
which, however, unless it is quite warm, it should not
be allowed to remain long in contact if its economical
preservation is of much moment, as this liquid will
when cold absorb a volume of the gas nearly equal to
its own bulk, though this may be readily recovered by
simply heating the water. But as for medical purposes
it is generally desirable to have protoxide of nitrogen
quite free from contaminations, it is necessary to adopt
such measures as will insure its purification, which may
be effected in the following manner: The retort containing the nitrate of ammonia, should preferably be of
glass, and be connected by means of glass or other
tubes of non-oxidizable matter joined by cement, India-
rubber, gutta-percha, or some similar substance, with
three glass vessels, such as Wolf's bottles, conjoined in
like manner. These last should be partially filled with
the following fluids, viz., the first one with a saturate
solution of protosulphate of iron, the second the same
of caustic potash or soda, and the third with plain
water. The connecting tubes, or those passing into
the purifiers on the side toward the retort should be
immersed into these respective liquids nearly to the

bottom of the vessel, but those on the side furthest therefrom should have their orifices above the liquid and close to the mouth of the bottle, which should be sufficiently capacious to admit of the insertion of a large stopper or cork containing the tubes of the greatest calibre that can be introduced. As nitrous oxide is generated it is compelled to thus pass through these different liquids which remove all adventitious matter and render it quite pure. This purified protoxide of nitrogen may then, as it comes over, be collected in suitable receivers, and be thus preserved for immediate use in its gaseous state, or, like carbonic acid, it may be introduced into some liquid which will take it up more or less freely, and for obvious reasons the fluid most convenient and useful for such purpose is water. Yet as water will not of itself absorb a sufficiently large quantity of nitrous oxide to render it very efficient for medical purposes, it is necessary to forcibly condense the gas therein, which should be continued until it is completely surcharged therewith, or to the extent at least of five volumes of the latter to one of the former, as it is not always desirable to administer much water, though frequently in this connection it offers no very serious objection, but, on the contrary, its free use is often therapeutically indicated, while at the same time it is always more or less essential as a diluent. The relative proportion of these two compounds will therefore vary according to the special necessities

of the case, the degree of admixture depending upon
the practical application, although the portability and
eligibility of this preparation will be measurably increased by the maximum quantity of gas to the minimum of liquid. In view of the fact, however, that
nitrous oxide gas will undergo liquefaction under a
comparatively moderate pressure, it is highly probable
that when thus forcibly introduced to any great extent
into such fluids as water, it may be condensed into a
liquid, and be mechanically associated if not chemically
combined therewith, as these two bodies seem to have
a strong affinity for each other, for I have known this
nitrous oxide water kept in an ordinary bottle with a
plain cork tied down, to preserve its active medicinal
properties for the long period of about seven years.
Protoxide of nitrogen may also be sometimes advantageously associated with alcoholic and other liquids for
therapeutic purposes.   But as the process for the separate liquefaction and solidification of nitrous oxide is
being rapidly perfected, the necessity for the primary
condensation with such fluids is diminishing; still as
the former is more difficult than the latter, and in consequence of the extremely important practical advant-
ages of this agent for medical purposes, its ready ac-
quisition is a great desideratum, it is hoped that both
methods will be so fully adopted as to promptly furnish
all that may be required.

V. Modes of Administration and Dose of Nitrous Oxide.

Protoxide of nitrogen may be administered with advantage either in its gaseous state by the lungs, or in association with liquids through this and other channels. Thus for instance, it may be introduced within the body in the aeriform state, separately or in conjunction with aqueous and other vapors, by direct inhalation or artificial insufflation through the air-passages, and in combination with water or other liquids through the alimentary canal by voluntary deglutition or the extraneous method of injection into the stomach and rectum. Also to some extent through the cutaneous surface. Either one or more of these modes of administration may be resorted to advantageously, according to the necessities of the particular condition requiring treatment, though in states of unconsciousness, and inability to inspire or swallow, as well as in cases of obstinate refusal to otherwise take it when needful, the artificial introduction of this agent may be the only alternative.

The effects of nitrous oxide, both in the form of gas by the pulmonary organs and in connection with liquids by the alimentary canal, are in general the same, though in the latter state and mode of exhibition, it sometimes seems to exert a more immediate and specific influence upon the chylopoietic viscera and genito-urinary organs, than when taken in the gaseous condition through the air-passages. In depressed cases, as of asphyxiation, suspended animation, or intoxication from various poisons, even when so profound as to temporarily place in abeyance the functions of deglutition, respiration, circulation, and innervation, the surcharged nitrous oxide water, and doubtless also other liquid preparations, may be injected into the bowels with such decided benefit as to sometimes induce speedy recovery, and always, so far as my experience goes, with a certain degree of advantage. Nitrous oxide may likewise be frequently administered with benefit in conjunction with wine and other alcoholic fluids.

This oxygenated liquid is a very convenient and valuable preparation, as it favors the more general distribution and application of its active ingredient, protoxide of nitrogen. When well made, nitrous oxide water is rather lively and sparkling, having a somewhat sweetish insipid taste, but with the addition of aromatic or other agreeable compatible substances it forms quite a palatable beverage. In view of this fact and the peculiarly pleasant as well as invigorant action of protoxide of nitrogen, this aqueous preparation will no doubt prove a valuable substitute for the various alcoholic and other objectionable compounds so freely employed for stimulating purposes. Hence in this connection, nitrous oxide water is of especial interest, as it will have a tendency to diminish the use of intoxicating liquors and the evils connected therewith. This oxygenated liquid will doubtless thus prove extremely useful in substituting a decidedly agreeable and highly sanitive for a positively injurious and destructive agent, as it will frequently promptly allay thirst and effectually satisfy the craving for some stimulus, without like alcohol either poisoning the body, stupefying the mind, or degrading the nature of those who use it. As before indicated, however, the salutary effects of nitrous oxide are not exclusively limited to the preceding, being moreover powerfully antagonistic to the deleterious effects of such agents, and thus quite active in the opposite direction, with perhaps one exception, for when taken too freely it is apt to unduly excite the sexual appetite, and thus indirectly encourage vicious practices, though even in this respect it is less obnoxious than alcoholic liquors, for unlike them it does not infuriate the passions and at the same time deaden the moral sensibilities, but while increasing erotic desire, also intensifies physical and mental energy, thereby augmenting instead of diminishing the power of resistance to immoral courses.

The *dose* of protoxide of nitrogen necessarily depends upon the varied circumstances of race, age, sex, constitution, temperament, idiosyncrasy, and special application, as well as the type, stage, intensity, and complications of the particular disorder requiring treatment. Thus, for instance, as a general rule, persons of a lymphatic can bear larger quantities than those of a sanguine or bilious temperament; and these latter more than the nervous,—these being frequently very susceptible to its excitant influence. Besides, as with most other agents, males will usually require more than females, adults than youths, and these again more than children. Hence, the quantity of nitrous oxide gas which may be exhibited at any one time, will vary from a very small to a comparatively large volume; as, for example, from one pint or less to several gallons, according to special indications and capacities; while that of the surcharged water ranges from one fluid drachm to one-half pint or more, repeated as often as may be necessary to produce the desired effect. The best plan, in ordinary cases, is to commence with a minimum, or what appears to be a suitable quantity for the case under treatment, and gradually or rapidly increase, until its action is manifest, or the required result is obtained, and continue according to circumstances, unless in conditions of extreme danger, when it may be freely employed without hesitation; for the necessity for stimulus is at such times generally so great, and the tolerance usually so much increased, as to render any apprehended evil from a liberal use of minor importance.

For the production of anæsthesia, however, some risk must always be run, as a comparatively large quantity will be required to be quickly administered; usually from two to four or more gallons of the gas, which should be repeated or its inhalation continued at short intervals, when a somewhat prolonged anæsthetic condition is required.

In thus enforcing the necessity for care in the use of nitrous oxide, it is not intended to excite undue apprehension respecting its dangerous tendencies, but merely to guard against overweening confidence in its harmlessness; for, though ordinarily quite safe, it is yet sufficiently potent to do great injury to the vital economy, even to the speedy destruction of life. Hence, under no circumstances should such an active agent be carelessly exhibited; for while in general the influence of protoxide of nitrogen is very agreeable, invigorating, and innoxious, yet, as it is powerfully stimulant with, as before mentioned, a special tendency to the blood, brain, nervous, vascular, and nutritive systems, as well as to the genito urinary organs, its indiscriminate and disproportionate employment is to be avoided, as it may prove injurious by superoxidation, overstimulation, and otherwise in the manner previously indicated.

Moreover, besides the immediate evils which may thus result from the rapid exhibition of a large quantity of nitrous oxide, there are others which follow its prolonged and inordinate use. Thus, if taken too freely for any length of time, it will cause undue molecular metamorphosis, with excessive disintegration of structure and emaciation of body, with abnormal excitability of kidneys, bowels, brain, and nervous system, unattended, however, unless in extreme cases, with any direct manifestation of weakness or debility, but rather with a restless energy, tending to vital exhaustion, from the superabundant consumption of the material elements and dynamic power of the organism. Nevertheless, though ordinarily thus objectionable in immoderate quantities, yet, when used discriminately, even to the extent of causing somewhat extreme disintegration of tissue and superexcitability, this agent will not usually produce so much waste or irritability of system as to prove injurious, but on the contrary by increasing the activity and augmenting the power of molecular and somatic life, be often very efficient in improving the tone of both body and mind. Hence, as before stated, in those cases of obesity, and other forms of abnormal nutrition, as well as in the many aberrations of function in which modified metamorphosis and increased disintegration with general invigoration is required, nitrous oxide offers, *cæteris paribus*, the most reliable means for the purpose, as it will not only reduce redundancy of flesh and regulate organic action, but will increase dynamization, and promote thereby the health and strength of the entire economy, both corporeally and psychically.

With regard to any absence of detail, or deficiency of cases in illustration of the correctness of the foregoing observations, it is proper to state, in conclusion, that in this paper I have mainly endeavored to give a very

general outline of the chemical constitution and properties, physiological influences, and modes of preparation, combination, and administration of protoxide of
nitrogen, and to point out some of its principal hygienic, therapeutic, revivifying, antidotal, and anæsthetic
applications, hoping at a future time to be able to
communicate more fully on the subject. Hence, notwithstanding the preceding remarks might be greatly
extended, I believe I have presented enough to exhibit
the superior sanitive power of nitrous oxide and demon-
strate its great practical value in the support of life
action. In consequence, therefore, of its pre-eminent
importance to the medical profession and humanity at
large, it is hoped these general observations will attract
the favorable attention of scientists to the transcendant
merits of this remarkable agent, and thereby aid in extending its application for the preservation of life,
diminution of suffering, and the promotion of health
and happiness.

www.ingramcontent.com/pod-product-compliance
Lightning Source LLC
Chambersburg PA
CBHW022007190326
41519CB00010B/1415